HOUSE OF LORE

# I Am the Solar System

REBECCA AND JAMES MCDONALD

I am the Solar System. I'm the place where your planet, Earth, and seven other planets circle around a very important star called the Sun.

I'm like a big neighborhood, and the Sun is right in the middle.

The Sun is the biggest object in the neighborhood. Its gravity keeps all kinds of objects circling around it. You can't see gravity, but it keeps things from floating away, even you!

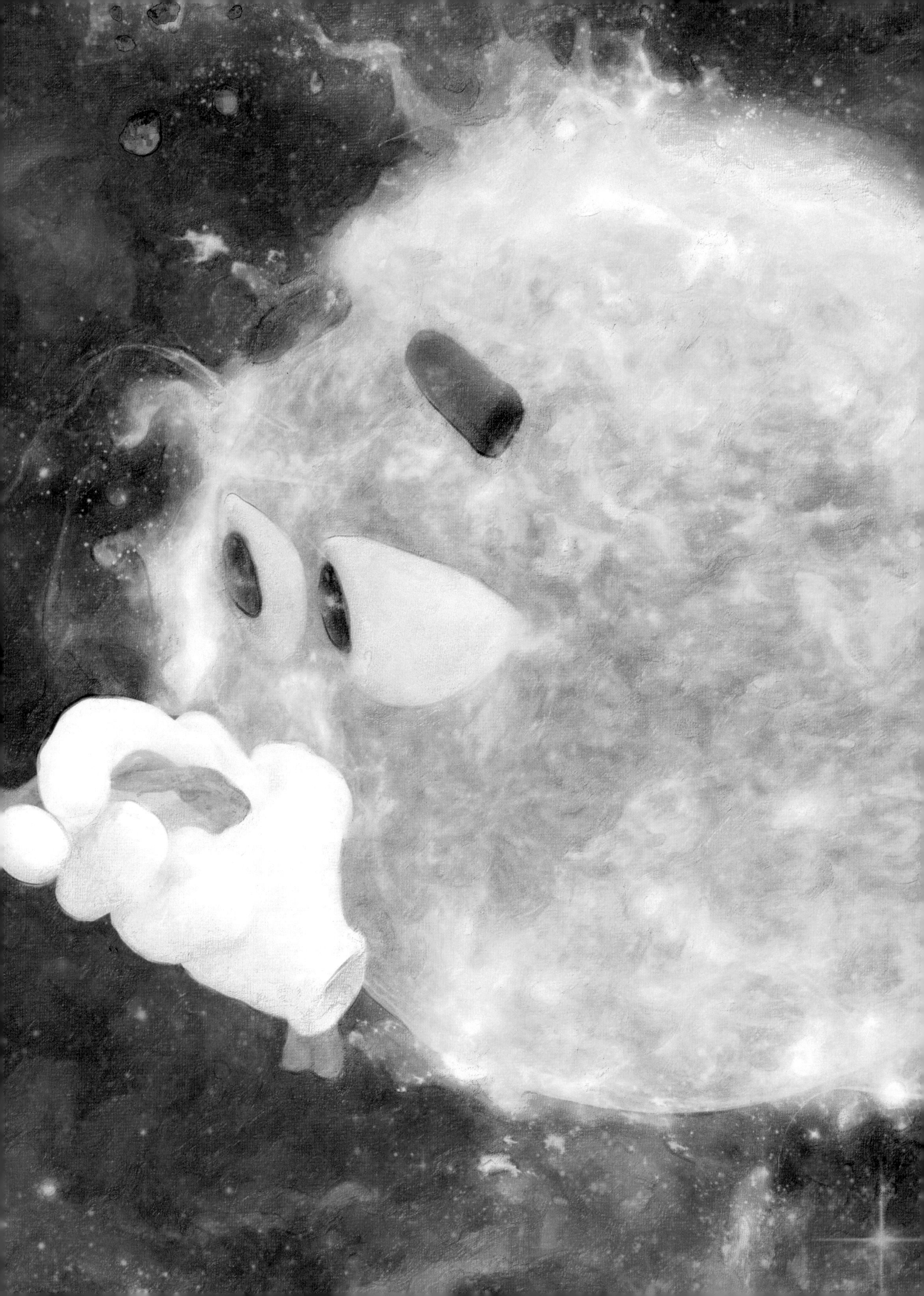

When one or more objects circle around another object, it's called orbiting. The planets orbit the Sun, and many have moons that orbit them.

Time to explore the neighborhood! There's no air to breathe once you leave Earth, and space is really cold, so a special spacecraft and spacesuit are needed for traveling the Solar System.

Ready for
blast-off...
3, 2, 1 !

Just ahead is the first and closest planet to the Sun, Mercury. It's also the smallest planet.

Venus is the second planet from the Sun. Even though it's not the closest, it's the hottest planet in this Solar System.

The third planet from the Sun is Earth. That's where you live! Earth is the only planet in this Solar System that holds all kinds of life on its surface.

The fourth planet orbiting the Sun is Mars. Many spacecraft have visited the red planet Mars.

Planets aren't the only things that orbit the Sun. Just after Mars is the asteroid belt. The asteroid belt is filled with dust, asteroids, and even a dwarf planet named Ceres.

Tighten your seatbelt, it could get a little bumpy, but don't worry about crashing into anything. Asteroids in the asteroid belt are far apart from each other.

Asteroids are mostly made of rock.
Some are tiny and some are gigantic!

Just past the asteroid belt is the fifth and largest planet orbiting the Sun, Jupiter! Scientists call it a Gas Giant, because its surface is made of gas and liquid. That means there's no ground for a spaceship to land on.

The sixth planet from the Sun is Saturn. It's also a Gas Giant. The rings that circle Saturn are made of ice, rock, and dust. Saturn's strong gravity keeps the rings from floating away.

The seventh planet from the Sun is Uranus. It's called an Ice Giant, because it's very icy and cold.

Neptune is the eighth and farthest planet from the Sun. It's also an Ice Giant. Scientists found Neptune by using numbers and math before they actually even saw it.

Neptune's the farthest planet from the Sun, but the Solar System doesn't end there! Next, is the Kuiper Belt. It's a giant freezing ring filled with rocks, dust, ice, and other space objects.

The dwarf planet Pluto is part of the Kuiper Belt!

Keep an eye out for bright objects with long streaking tails speeding through space. They're called comets.

The Sun holds the planets and other space objects in place with its gravity. It also protects the Solar System in a giant bubble made of solar wind.

The solar wind comes off of the Sun's hot surface and blows in every direction, past the planets, even beyond the Kuiper Belt! It blows as far as it can, until it forms a giant bubble around the Solar System called the heliosphere.

Outside of the heliosphere is where scientists say the Solar System comes to an end, and interstellar space begins. Interstellar space is the space between stars.

There are only two spacecraft that have made it out of the Solar System and into interstellar space, Voyager 1 and Voyager 2, and they are still exploring!

I'm not the only Solar System in space. There are many more stars out there with planets that orbit them, but so far, I'm the only solar neighborhood with a planet that's filled with life.

Which planet in the Solar System is the only planet known to have life?

How many planets are in your Solar System?

What are the names of the 2 belts in the Solar System?

What is an object doing when it circles around another object?

What holds the planets and other space objects in place and keeps them orbiting the Sun?

When the solar wind forms a giant bubble around the Solar System, what do scientists call this bubble?

I Am the Solar System



ISBN: 978-1-950553-23-5

First House of Lore paperback edition, 2020

Visit us at www.HouseOfLore.net

CHECK OUT THESE OTHER TITLES FROM HOUSE OF LORE
I Am Earth
Rebecca and James McDonald
I Am the Solar System
REBECCA AND JAMES MCDONALD
I Am the Moon
REBECCA AND JAMES MCDONALD
I Am the Sun
REBECCA AND JAMES MCDONALD
I Am a Bee
I Am a Frog
I Am a Dinosaur
I Am Tyrannosaurus Rex
I Am the Alphabet
REBECCA AND JAMES MCDONALD
Planting ABC in a Garden of Memory
Rebecca and James McDonald
ALPHA THE ALPHA-BOT GUARDIAN OF THE ALFURBETS
REBECCA AND JAMES MCDONALD
THE ALPHABET BY COCKROACHES
REBECCA AND JAMES MCDONALD
The Scribbles
REBECCA AND JAMES MCDONALD
Rainy Day Poems
If You Don't Understand, Raise Your Hand
Bo the Bear BUILDS a Monster Truck
REBECCA AND JAMES MCDONALD
Why Mama Why
Rebecca and James
WILFORD AND BLUE The Kite Calamity
REBECCA AND JAMES MCDONALD
WILFORD AND BLUE Mud Madness
REBECCA AND JAMES MCDONALD